- **Name:**

  Eng./Mostafa   Yacoub   Abdellatif

  Mahmoud

- **Nationality: Egyptian**

- ORCID: 0000-0002-9991-4624

- Email: moshhaabma2015@gmail.com

- Qualification: civil engineer Cairo

  University 2003

Proof for infinite prime numbers:

- **In this paper or research, we will make a proof for infinite prime numbers based on my discovered formula that connects prime and composite numbers.**

- **My discovered formula:**
  Definitions:

- **Array PTBP**

- **It is the following Array of odd numbers**

$$\begin{vmatrix} 1 & 3 & 7 & 9 \\ 11 & 13 & 17 & 19 \\ 21 & 23 & 27 & 29 \\ 31 & 33 & 37 & 39 \\ 41 & 43 & 47 & 49 \\ 51 & 53 & 57 & 59 \end{vmatrix}$$

And so on....

- For a given set of consecutive prime numbers whose numbers = n that start with prime number 3 and end with prime number F and not including prime number 2 and prime number 5

- i.e.

  set=[3,7,11,13,..............................

  ....,F]

- S=product of those consecutive

  prime numbers

  i.e

  $$S = \prod_{i=3}^{i=F} (i)$$

- Range=$R_k = 10 \times S \times k$

- Where k = [1, 2, 3, 4, ..., ∞(infinity)

  i.e

  $R_1 = 10 \times S \times 1$

  and

  $R_2 = 10 \times S \times 2$

  And so on

- **Number of composite numbers that belong to Array PTBP and created by the effect of those consecutive prime numbers within the range R_K**

- $= [(K \times 4^{\times \frac{S}{3}}) + ($

$$\sum_{j=7}^{j=F} (K \times 4 \times (\frac{S}{j}) \times$$

$i = prime\ number\ befor\ current\ prime\ number\ j$

$$\prod_{i=7} \qquad (\frac{i-1}{i})$$

)

]-(n)

- **Where j =consecutive values of prime numbers**

  7, 11, 13,..............., F

- **And i= consecutive values of prime numbers**

  3, 7, 11, 13,........, prime number before current j prime number

- The previous formula can be applied for any number of consecutive prime numbers that start with prime number 3.

- The first term $(k \times 4 \times \frac{S}{3})$ represents the count of unique Composite numbers +1 that belong to the Array PTBP and are created by prime number 3 within the range

$R_k = 10 \times S \times k$

- **The second term**

$$\sum_{j=7}^{j=F} \left(K \times 4 \times \left(\frac{S}{j}\right) \times \right.$$

$i = prime\ number\ befor\ current\ prime\ number\ j$

$$\prod_{i=7} \qquad \left(\frac{i-1}{i}\right)$$

Represent the count of unique

Composite numbers+n-1 that belong

to the Array PTBP and are created by

each prime number after the prime

number 3 within the range

$R_k = 10 \times S \times k$

- **The third term (-n)**

  Subtracting n (number of consecutive prime numbers starting from prime number 3) because the count of composite numbers generated from those consecutive prime numbers includes the count of those prime numbers in the range

  $R_k = 10 \times S \times k$

- Explanation and proof for my theory in my previous paper (prime number theory)

- We will mention only the concept of number cycle

- We can use the number cycle concept to understand the behavior of consecutive primes in creating composite numbers.

- i.e.

$$S = \prod_{i=3}^{i=F} (i)$$

Range=cycle range= $R_k = 10 \times S \times k$

Where k= [1, 2, 3, 4, ...,

∞(infinity)

i.e.

R_1=10 x S x 1

R_2=10 x S x 2

And so on

- Now consider only one k value =1

- For a set of consecutive primes and according to my formula the result will be a certain number

  = V less than 4 x S

  $V < 4 \times S$

  Then $(4 \times S) - V = W$

- **The next prime number after prime number F (prime number G) creates some composite numbers including the count of that prime number within range**

$$= R_1 = 10^{\times \prod_{i=3}^{i=G} (i) \times} 1$$

Count of composite numbers

$$= \left\{ \left(1 \times 4 \times \left(\frac{\prod\limits_{i=3}^{i=G} (i)}{G} \right) \times \prod\limits_{i=3}^{i=F} \left(\frac{i-1}{i} \right) \right) \right\}$$

$$= \left(4 \times \prod\limits_{i=3}^{i=F} (i-1) \right) = W$$

So each cycle corresponding to a certain set of consecutive primes numbers connected to the next cycle that produced by adding the next prime number to that set so that the new prime number added to the set must produce account = W

- W = the complementary number to the previous cycle within range R = range of the previous cycle multiplied by G

- i.e

$$10^{\times \prod\limits_{i=3}^{i=G} (i)} = 10^{\times G \times \prod\limits_{i=3}^{i=F} (i)}$$

- **And for more explanation**

  Prime number 3 has a cycle range

  $$= 3 \times 10 = 30$$

  And produce some composite numbers = 4 In this first cycle including the count of prime number 3 itself

  So we have S = 3

  And we have $4 \times 3 = 12$ in that cycle

- **The reaming = $4 \times 3 - 4 = 12 - 4 = 8$**

  So the next cycle is produced by prime numbers 3, 7

  With S= $3 \times 7 = 21$

- **So prime number 7 must produce 8 numbers including prime number 7 itself within 7 cycles of the previous prime numbers which has only prime number 3**

- **And now we have**

  $$4 \times 3 \times 7 - 4 \times 7 - 8 = 84$$

- **So prime number 11 must produce 48 numbers including prime number 11 itself with 11 cycles corresponding to prime numbers 3 , 7**

  And so on

- I.e. we always have new numbers that are not produced by our defined set of consecutive prime numbers

- So number of prime numbers

  $= \infty$ (infinity)

www.ingramcontent.com/pod-product-compliance
Lightning Source LLC
Chambersburg PA
CBHW071218290526
45796CB00008B/290